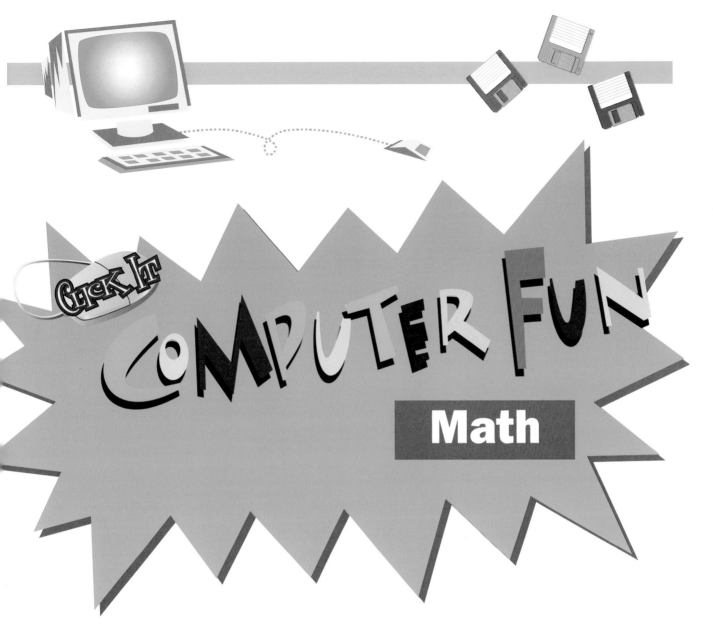

By Lisa Trumbauer

The Millbrook Press
Brookfield, Connecticut

Copyright © 1999 by The Millbrook Press, Inc. All rights reserved.
Published by The Millbrook Press, Inc., 2 Old New Milford Road, Brookfield, CT 06804.
Printed in the United States of America.

Produced by 17th Street Productions,
a division of Daniel Weiss Associates, Inc.
33 West 17th Street, New York, NY 10011

Editor, Liesa Abrams
Special Projects Editor, Laura Burns
Cover illustration by Sam Ward
Interior design and illustrations by Sydney Wright

Library of Congress Cataloging-in-Publication Data

Trumbauer, Lisa 1963-
    Click it. Computer fun math / by Lisa Trumbauer ; illustrated by Sydney Wright.
        p.   cm.
    Summary: Teaches facts and skills about mathematics, while helping the reader to become comfortable with using a computer.
    ISBN 0-7613-1504-7 (lib. bdg.). — ISBN 0-7613-0996-9 (pbk.)
    1. Computers—Juvenile literature. 2. Mathematics—Data processing—Juvenile literature.
[1. Computers. 2. Mathematics—Data processing.] I. Wright, Sydney, 1957- . II. Title.
QA76.23.T78   1999
510' .285—dc21   98-37602
                                                                                      CIP
                                                                                      AC

1 3 5 7 9 10 8 6 4 2

Contents

Introduction 4

CHAPTER ONE: Get in Shape
Shaping Up 9
Modern Madness 11
It's About Time 12

CHAPTER TWO: Number Invasion
Picture-Perfect 14
Wild Cards 16
Brain Puzzlers 17
Fraction Action 19
What Do You Think? 20

CHAPTER THREE: Measuring Up
Sky High 22
You Rule! 24

CHAPTER FOUR: Money Matters
Making Money 26
Shop-o-rama 27
Shopping Spree! 29

CHAPTER FIVE: Sports Zone
Win, Lose, and Draw 30
High Scoring 31

Now What?!

You'll be surprised to find out that you can have lots of fun on your computer and—at the same time—learn all sorts of things about math. The activities in this book add four things together: computers, math, fun—and you!

Why Math?

When you watch a favorite sport or game, how do you know who's winning? By adding up the baskets or goals or touchdowns, right? Or how about when you go to the movies? How do you know how much money you need for tickets and popcorn and other stuff? By adding up prices. I bet you like to know how tall you are too, and how to tell when you've grown a couple inches. In case you haven't already figured it out, finding all of these answers takes math skills! Tons of other things you do involve numbers or shapes also, which means—yup—more math!

But since this is a computer book (duh—you knew that!), you have to know some of the basics first. Here's your computer, inside and out:

What it is: You know that big screen that looks like a TV? That's the **monitor**.
What it does: It shows you what you're working on.

What it is: See all the buttons with letters and numbers on them? They're on the **keyboard**.

What it does: Here's where you type in what you want the computer to do and also all the words you want to appear on the screen.

What it is: Can you find the funky-looking curvy tool with one or two big buttons on it? That's the **mouse**.

What it does: It lets you move around the computer screen and choose where you want to go. How? Simple—once you understand some mouse lingo:

Cursor: This shows you where you are on the computer screen. Depending on which program you're using, it may look like a blinking line, an arrow, or an icon. In the **Paint** program, you can move the cursor by moving the mouse.

Click: When the instructions tell you to "click," you push the left button on the mouse and then release it quickly.

Drag: When you need to "drag" the cursor across the screen, you hold down the left button on the mouse and move it across the **mouse pad** it's resting on.

Write On

For the activities in this book, you'll need to understand how to use the **word-processing** program. That's what you use for writing, like when you want to write notes to a friend or make up a story about all your wild adventures as captain of a spaceship. Many different word-processing programs are available, but one of the most popular is **Microsoft Works**. The following activities are based on this program, but you can do them with any other program that your computer has. Here are some tips on how to use **Microsoft Works**:

Click on the word **Start** at the bottom of your computer screen. See that list of words above it? Move your mouse up until the word **Programs** is highlighted, then move your mouse to the right and you'll see the names of all your computer's programs. Look for **Microsoft Works**. Found it? Great! Click on it once, and then you'll see three choices. Click **Works Tools** once and then **Word Processor** once. Easy enough, right? Now you should have a blank screen, just waiting for you to fill it up with your writing!

At the top of the screen you'll see some words. This is called the **Menu Bar**. Move your mouse to one of these words, and an arrow will appear. Click, and you'll get a list with more words, called **commands**. You use them to tell your computer what to do. Here are the commands you will use:

- In **File:** Save, Page Setup, and Print
- In **Edit:** Cut, Copy, and Paste
- In **Insert:** ClipArt, Drawing, and Object
- In **Format:** Font and Style

Below you will see a row of small pictures. This is the **Tool Bar**. The Tool Bar lets you do some things without using a command. For example:

- **Font Box:** This is the first thing you see on the left. It tells you the name of the type style, or font, that you are using. You can choose a new font (**FONT**, font, font, *font*) by clicking the arrow to see your choices, then clicking on the one you want.

- **Numbers:** These numbers tell you the size of your type. Click on the arrow to see how large or small you can make the type—small, normal, huge!

- **B I U:** These three boxes let you change the way the type looks.

    B stands for boldface. It makes the type **darker**.
    I stands for italics. It makes the type *slanted*.
    U stands for underline. It draws a line under the type.

- **Lines:** You might see three or four boxes with straight lines in them. These let you move the words you type to different parts of the screen. You can choose to put them all the way to the left, in the middle (centered), all the way to the right, or all lined up on both sides (justified).

Before you begin to type, choose a font and a size. You can do this on the Tool Bar by clicking a new font or type size. You can also do this by clicking **Format**, then clicking **Font and Style**. Here you will see fonts and sizes and **colors!** Choose a color, just as you would a font or a size—by clicking the arrow beside the **Color Box** to see the choices, then clicking the color once. When you've made all your choices, click the **OK** button to return to the main screen.

Paint Up a Storm

So now you're an expert at the writing stuff. What else is there? Painting! Did you know you could be a master artist without picking up a single paintbrush? You can actually paint pictures on your computer! Since you'll be doing a lot of that in the activities in this book, here's a guide to your **Paint** program:

Click on the word **Start** at the bottom of your screen. Remember this list? Click on **Programs** again, but this time choose the word **Accessories**. Yes, there's *another* list of choices. See the word **Paint**? You guessed it—that's where you click!

Whoa! Look at all your paint tools! And colors! Here's what some of the tools in the **Tool Box** can do:
- Pencil: draws a line
- Paintbrush: paints a thicker line
- Paint Can: fills an object with color
- Spray-Paint Can: makes splotchy, star bursts of color
- "A" Icon: makes a box for you to type in
- Shapes: these make exact shapes
- Dotted-Line Box: can move or delete art or type
- Eraser: erases color

To change a color, just click the color you want in the **Paint Box**. If you click the **Paint Box** twice, you'll get a grid with even more colors!

HINT!

Okay, we know the pictures you paint will be totally amazing, but just in case you want to sneak in some of the stuff by the pros, here's how to do it: ClipArt. These are pictures that have already been drawn and are stored inside your computer. In your word-processing program click Insert, then ClipArt. You'll see pictures of all kinds of things, from flowers to people to computers! Click the arrows to move the list up and down to see them. (This is called scrolling.) Then click once on the picture you want. A box will appear around it. Click the Insert button. The picture will appear on your word-processing page! The clip art in this book comes from sources other than Microsoft Works, and you can always buy software that contains more clip art!

PSST!
Exit the Paint program by clicking the box with the X in the upper-right corner.

Now that you know how to write and draw on the computer, there are only two more things to learn before you get started on the *fun* part—the activities! What good would your work be if you couldn't show it off to people? How do you do it? Easy—**Save** and **Print**.

Saving

Click on the word **File** at the top of the screen, then click **Save**. A box called **Save As** will appear. This box may have some folders in it. **Folders** are where the documents you save are kept. Choose a folder to keep your work in, or create a new folder. At the top of the **Save As** box, you'll see a folder with a star beside it. Click it once. In the box that appears, type the name you want to use for your folder over the highlighted words New Folder. Then hit the **Return** key twice, which will open your new folder. Type the name of your document in the white box that says **File Name**. Click the Save button. Your work has been saved!

Once you've saved your file, you'll want to print it out (see below) or start a new one. Click into **File**, then **New** to start a new page. Your old art will disappear, and a new, clean canvas will take its place. To open the file again, click **File**, then **Open**. Click on the name of your file once, and click the **Open** button again. Your work will appear!

Now Print!

You'll have to print out your "paintings" to put together mobiles, posters, and other projects. It would be best if you had a color printer. If not, don't worry!

1. Click **File**, then click **Print**.

2. A box with print choices will come on the screen. Some of the activities in this book work better if you choose to print your page in the **Landscape** format. When you need to do that, the instructions will explain how. Otherwise just click **OK**, and your page will print.

That Should Do It!

The activities in this book are based on Windows 95, using **Microsoft Works** and the **Paint** program. Many versions of **Microsoft Works** exist. So you may have to alter the instructions slightly to fit your computer. Also, the pictures show how a finished product *may* look. Don't worry if your art looks a bit different. It probably will! That's because you used *your* ideas and *your* computer!

Computers can do about a billion things, so get ready to see all the fun stuff you can do with your computer. Experiment by playing with the pictures, by cutting and pasting, and by choosing fonts, sizes, and colors. Be crazy, be creative, and have a blast!

Chapter One

Get in Shape

You probably had to learn the names of all the shapes when you were younger. That's because shapes are important in *geometry*, a major math skill. There are all kinds of rules about shapes, like how the four sides of a square have to be exactly the same size. Plus you can figure out lots of number problems just by knowing how much space is inside a circle. For now, experiment more with shapes while you learn all about your Paint program!

Shaping Up

Okay, so you know the deal—triangles, squares, circles . . . old news, right? Well, check out all these funky ways to make them, along with other kinds of shapes. Then you can even put your shapes together for a mobile you can hang in your room!

Steps:

1 Go into the **Paint** program.

2 Click the **Rectangle** tool, then move the cursor onto the blank canvas. Press and hold down the mouse button, then move the mouse to drag the cursor. A rectangle will form! Release the mouse button when your shape is ready. Practice making squares and rectangles of all sizes.

3 Click the **Ellipse** tool, which makes circles and ovals. Again, move the cursor onto the page, press and hold down the mouse button, then drag it to make the shape. When you like your shape, release the mouse button. Practice making all kinds of circles and ovals in different sizes.

4 Now fill your shapes with color. Click a color in the **Paint Box**, then click the **Paint Can**. Click inside the shape. It's filled with color, right?

PSST! If you can't fit all your shapes on one canvas, just save what you've done, and open up a new canvas for the rest of your shapes!

5 You can draw other shapes with the **Straight-Line** tool. Try drawing a triangle, a star, a parallelogram, and a hexagon. When you fill them with color, *be careful!* Make sure all the lines are connected (use the **Magnifier** tool). If they're not, your entire page will fill with color!

HINT! If you accidently fill your page with color, don't panic! Click Edit, then click Undo. The color will disappear!

6 Also try the **Pencil** and **Paintbrush** tools to draw shapes. The **Pencil** gives you a thin line, and the **Paintbrush** gives you a thicker line. Fill the shapes with color, but again make sure the shape is closed up so you don't color the whole page!

7 On a new canvas, try making pictures with your shapes! For example, a rectangle could be an apartment building, and squares could be the windows. A big circle could be a wheel. A square with a triangle on top can make a house. Make as many things as you can with the shapes you draw!

8 Don't forget to **Save** and **Print** your paintings. Before you print them, click **File**, then **Page Setup**. You'll see a box called **Orientation**, with two choices—**Portrait** and **Landscape**. Click **Landscape**, then click **OK**. Now print!

COOL IDEA! Cut out all your shapes, and tape each to some string. Tie the string to a hanger, a paper plate, or a stick to make a mobile!

Modern Madness

Have you ever been to a museum and seen paintings with lots of shapes and colors splashed all over them? These are abstract paintings, and a lot of modern artists paint them. The artists paint what they feel—but they're also using math because recognizing shapes is a major math skill, as you already know. Make your own modern-art painting, using all the colors and shapes you want.

Steps:

1 Go into the **Paint** program. This time, click on the funny looking **Polygon** tool.

2 Hold down the mouse button, drag the mouse to move the cursor across the screen, and then lift your finger. Keep doing this—drag and lift, drag and lift.

3 Experiment and have fun! Make the lines cross one another, form triangles and squares inside others, or whatever you want. Try to make the most fantastic picture you can. Make sure when you are finished that the last line connects to another line. This closes the shape.

HINT! The Paint Can can change colors too. If you don't like a color, click a new one. Click the Paint Can on the old color to change it!

4 Now fill in each part of the big shape with color. Click a color in the **Paint Box,** and click the **Paint Can.** Pick a part of your shape, and click into it. Color all the parts of the shape this way.

5 Save your art and **Print** it out. Make sure the page is set for **Landscape.**

COOL IDEA!
Make a frame for your artwork! Measure and cut strips of brown construction paper. Glue the strips around your paper. Hang your modern art in your room!

It's About Time

Telling time is a skill that involves tons of math. First of all, you need to know how there are sixty seconds in a minute, sixty minutes in an hour, and twenty-four hours in a day. Then you'll need addition and subtraction to work out how much time you have to get things done or how long you have until you need to be somewhere. Plus a clock is also a shape, so when you draw a clock you're learning about numbers and shapes at the same *time!*

Steps:

1 Go into the **Paint** program, and make a big circle using the **Ellipse** tool.

2 Then add numbers to the clock. Click the **A icon.** Hold down the mouse button and drag it to form a text box. Type in the number 12.

3 Click the **Dotted-Line Box** tool. Place the cursor to the upper-left side of the number, then hold down the mouse button and drag it over the number. You will have made a dotted-line box around it. When you place the cursor back inside the box, you should see a large, thick cross. Hold down the mouse button, and drag the mouse.

PSST!
The printouts will be smaller than the pictures on your screen, so be sure to make the clock face and hands as big as possible!

4 The number moves! Move the number to its correct place on the clock. Follow steps 2 and 3 to make the other clock numbers, moving each one into position. **Save** and **Print** the clock face. Cut it out of the paper.

5 Then make the hands for the clock on a new screen. Remember to make a small hand to point to what hour it is, and then a bigger hand to point to how many minutes have passed in that hour. Draw the hands with either the **Pencil**, the **Paint Brush**, or the **Shape** tools. Add color to the hands too, with the **Paint Can**.

6 **Save** and **Print** the hands, and then cut them out. Put the clock together by attaching the hands in the middle of the face with a brass fastener, which you can get from your parents. Hang the clock in your room. The next time you have to be somewhere important, set the clock to that time to help you remember it!

The Dotted-Line Box tool can also be used to erase, or delete. Click the tool and drag a box around an object. Then hit the Delete key or the Backspace key. Now you see it, now you don't!

Chapter Two

NUMBER INVASION

Now that you've experimented with shapes, it's time to get back to numbers. Number knowledge is superimportant for counting and adding, things you probably do in school. That's not all numbers are used for, though. If you check out stuff around your house, you'll probably see numbers on just about everything—books, TV sets, food labels, even video games. Here's a chance to sharpen your number skills on the computer.

PICTURE-PERFECT

What do pictures have to do with math? This book isn't about art; it's about *math!* In this activity you choose any picture you want and then print out tons of copies to use as fun tools for practicing addition, subtraction, and even multiplication and division!

Steps:

1 Go into the **word-processing** program. Look at your **ClipArt** (under **Insert**, remember) and choose a picture you like. Insert it onto your page.

2 Click **Edit**, and then click **Copy**. It will seem like nothing has happened, but the computer has actually made a copy of your picture. Click outside the picture, so you see a big, blinking cursor next to it. Click **Edit**, then click **Paste**. See, there's another picture!

3 Click **Edit**, then **Paste** ten more times, and you'll see ten more pictures. You already had two, so now you have a total of twelve. Print out the pictures, then cut them up so each one is separate.

4 First use your pictures to practice adding. Grab some of the pictures, count them, and put them in a pile. Take a few more pictures, count them, and put them in another pile. Add the two numbers together to see how many pictures you would have if you combined the piles. Check your answer by combining the two piles of pictures and counting them!

5 Now try subtraction. Put the pictures back into their two piles. Subtract the number of pictures in one pile from the total number of pictures you added together in step 4. Your answer should be the number of pictures in the other pile!

+

―――――――

12

6 For some clues on division and multiplication, put all twelve of your pictures together, then separate them into two equal piles of pictures. Count the pictures to make sure each pile has the same amount. Now you have two piles of six pictures, right? So you can see that when you divide twelve by two (which is what you did when you put twelve pictures into two different piles!), the answer is six.

7 Multiplication is the opposite of division, so this time you're putting the piles back together. And two piles of six pictures equals twelve pictures—so now you know that six multiplied by two (that means two piles of six) equals twelve!

Keep playing with your pictures to get more practice with addition, subtraction, multiplication, and division.

MATH WILD CARDS

Have you ever played a fun card game, like Go Fish, Old Maid, or Crazy Eights? Now you can make your own cards and use them to invent math card games!

Steps:

1 Go into the **Paint** program. Click on the **Rectangle** tool, and draw the outline of a big rectangle for your card.

2 Inside the card draw the number 1. Experiment with the **Shape** tools, the **Pencil**, and the **Paintbrush** to draw it. If you want, fill the number with color too. Then draw one picture of something on the card, like one flower or one tree.

3 **Print** a few copies of the number card.

4 Now, with the **Eraser** tool, erase the number inside the card, and make another one, following step 2. This time, draw the number 2 and paint two pictures of something, such as two fish or two ice-cream cones. **Print** it out. Follow these steps for all the numbers up to 10. Next make cards for math operations, like a plus sign, minus sign, multiplication and division signs, and an equals sign.

PSST!
Don't forget that the printouts will be smaller than what's on your screen!

5 Once all of your cards are printed out, use them to work on the skills you practiced in "Picture-Perfect" by mixing and matching number cards with math operations cards and figuring out the answers.

Then print out more copies of the cards and make up some games to play with your friends.

Real decks of cards always have *joker* cards that are used as wild cards in certain games. Make some wild cards with fun, crazy pictures you paint and then put them in your number card deck!

 **BRAIN PUZZLERS**

Have you ever played those puzzle games where you have to match up questions and answers? Some of them can be really tricky. In this activity you get to make your own puzzle game! Making these math puzzles on your computer will be a great review for what you've done so far, and you can design them however you want.

Steps:

1 Go into the **Paint** program. Click the **Rectangle** tool, and make a medium-size shape.

2 Now click the **Pencil** tool. Draw a zigzag line down the middle of the rectangle.

3 Choose a color in the **Paint Box**, then click on the **Paint Can**. Fill one half of the rectangle with color. Choose a second color, and fill the other half with that color.

4 With the **A icon**, type in a subtraction equation, like 29-6, on the left side. The equation is the problem you're trying to solve, or the question you need an answer to. Then type the answer on the right side. Type 23 for 29-6, since that's the answer to that equation!

5 Now **Save** and **Print**. Make other math equations, following steps 1 through 4. Use your pictures and your number cards for help figuring out equations and answers.

COOL iDEA!
Time yourself matching all the puzzle pieces. Then time a friend, a parent, or another family member. Who has the fastest time?

6 To make the equations into puzzles, glue them to cardboard and cut them out. Cut apart the equations from the answers.

7 Mix up all your puzzle pieces. Now try to pass the test! Put the math-problem puzzles together by matching the equations with the correct answers.

FRACTION ACTION

Have you ever eaten half a pizza? Or a piece of cake that's been cut into four slices? Besides making yourself really stuffed, you're also doing something else—using fractions! Fractions are an important math skill, so here's how to learn all about them while making some fun food pictures.

Steps:

1 Go into the **Paint** program. With the **Ellipse** tool draw a circle. Imagine it's a pizza, and add colors, like yellow cheese, red tomato sauce, and then your favorite toppings, like pink circles of pepperoni or green strips of green peppers. This is a whole pizza pie. In fraction language, a whole of something is the number 1. With the **A icon**, write the number 1 below the pizza.

2 Now **Copy** and **Paste** the pizza to make a second one. Click on the **Dotted-Line Box** tool and draw a box around your pizza, like you did in "It's About Time," on page 12. Then click **Copy**, under **File**. Click next to your first pizza. Then click **Paste**. Move your new pizza to the side of the first one. Then draw a line down its middle. You've cut the pizza in *half*. How many pieces do you have? Two! So two halves make a whole. Write:

$$\frac{1}{2} + \frac{1}{2} = \frac{2}{2} \text{ or } 1$$

beneath your pizza. You've just written a fraction equation.

3 Now **Paste** the pizza again to make a third one. Draw a line across the middle and another one down the middle. Now you've cut the pizza into *quarters*. How many pieces do you have now? Four! So four quarters make a whole. Write another fraction equation beneath the pizza:

$$\frac{1}{4} + \frac{1}{4} + \frac{1}{4} + \frac{1}{4} = \frac{4}{4} \text{ or } 1$$

4 Do you see the pattern? You add the top numbers of the fractions to get the total fraction. If the top number and the bottom number are the same, then you have a whole, or 1!

5 Make another pizza, and slice it up in other ways. Try to slice it into three equal pieces. These are thirds ($\frac{1}{3} + \frac{1}{3} + \frac{1}{3} = \frac{3}{3}$ or 1). Or slice it into six equal pieces. These are sixths ($\frac{1}{6} + \frac{1}{6} + \frac{1}{6} + \frac{1}{6} + \frac{1}{6} + \frac{1}{6} = \frac{6}{6}$ or 1).

6 **Save** and **Print** out your pizzas. Glue them to construction paper to keep as a guide to help with your fraction action.

WHAT DO YOU THINK?

Which pizza topping do your friends like better—sausage or pepperoni? Which CD-ROM game do they think is the most fun to play? Which ice-cream flavor rates the highest? These are things you would ask in a *survey*—a list of questions that ask for people's opinions on all sorts of stuff. To figure out the results of a survey, you have to add up the answers, and that means using math. Make up your own survey and then learn how to make a *bar graph* and a *pie chart* to show the answers you get!

Steps:

1 First, come up with a question you want to ask your friends. Decide how many things they'll get to choose between, like three different types of soda (*Coke*, *Sprite*, or *Mountain Dew*). Now take the survey and write down how many people choose each kind as their favorite.

2 Time to make the bar graph! Go into the **Paint** program. With the **Straight-Line** tool draw a line across the bottom of the canvas. Then draw another line going up the screen from the left corner of the first line. The result should look like a giant L on your canvas.

3 Underneath the first line you drew use the **A icon** to write the names of the choices you gave your friends. For example, if you asked about soda, you would write the word *Coke*, then leave some space, then write the word *Sprite*, then leave more space, and then *Mountain Dew*.

4 Now look at the answers you got from your friends. How many people chose each thing? Using the **Rectangle** tool, draw that many boxes over the word. If seven people chose Coke, draw seven boxes above the word *Coke*. Using the **Paint Can**, fill in your boxes with color. Do the same for each choice. **Save** and **Print** your bar graph. Now you can tell which was the most popular, by looking at the column with the highest number of boxes.

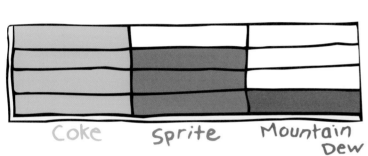

5 Next show your results in a pie graph! With the **Ellipse** tool make a circle. Slice the pie into the same number of slices as the number of people you surveyed. If you talked to ten friends, cut the pie into ten slices. Choose one color for each answer, then color in the right number of slices with the **Paint Can**. For instance, if four people chose Coke in the soda question, color in four slices with the color you chose for Coke.

The largest colored chunk of pie is the most popular answer!

PSST!

Remember to make sure the lines of the slices are connected so you don't fill in too many slices at once!

Chapter Three

Measuring Up

Do you ever wonder how much a human head weighs, or how tall the Empire State Building is? Want to know how long your dad's legs are? How about your mom's? You can find the answers to all of these questions by measuring, which is—you know what's coming—another math skill!

MATH Sky-High

Isn't it great the way you get taller all the time? Don't you want to know whenever you've grown another inch? If you want to see how much you're growing, make this wall chart you can use to measure yourself!

Steps:

1 Go into the **word-processing** program. Is there a ruler at the top of your page? If not, click **View**, then **Ruler**. The ruler should measure 6 inches long. (If it doesn't, click **File**, then **Page Setup**. Change the right and left **Margins** to 1.25 inches. Click **OK**.) If there is still a **Help Menu** on your screen, you need to make it smaller by hitting the X in the upper-right corner of the menu.

2 Click into the **Header**. Hit the **Tab** or the **Space Bar** key until the cursor appears under the number 1 on the ruler. Type in the number 1. Continue for the numbers 2, 3, 4, and 5, placing the numbers under the ruler numbers. Make your type big if you like. Now you have the inches marked on your screen.

3 Now add color. Click onto the main part of the page. Click **Insert**, then **Drawing**. A paint canvas will appear. It looks a little like the paint canvas in the **Paint** program, but it doesn't have as many tools. Click a color in the **Paint Box**. Click both **Line** and **Fill**. Choose the **Rectangle** tool, and draw one on the canvas. It will fill with color.

4 Click **Edit**, then click **Select All**. Click **Edit** again, then click **Copy**. **Exit** the drawing tool by clicking the **X** in the upper-right corner of the paint canvas. A message may ask if you want to save your drawing. That's up to you!

5 Back on your word-processing page, click into **Edit**, then **Paste**. Your colored box will appear! Make it fit across the entire page. Click it once so the frame shows up. Then drag the **Resize** dots.

6 **Save** and **Print** it out.

7 Make more sections for the wall chart, each time increasing the numbers. So for the second page type number 6 in place of the 1, number 7 in place of the 2 on the ruler, and keep going like this on every page, printing out each one, until you have enough sheets to measure 45 inches. Change the colors of the pages if you want, and paint pictures in the main rectangle too.

8 Glue or tape all the pieces together in number order. You might have to cut some of the white paper off so the numbers match up evenly. Hang your measuring chart on your bedroom wall to mark how tall you grow!

MATH — You Rule!

You measure something to see how big it is, but also so you can compare it to other stuff. Like, which is longer—your pet lizard or your friend's hamster? Create a ruler on the computer to measure and compare stuff around your home. (Now you can find out who really has the biggest bedroom!)

Steps:

1 Go into the **word-processing** program. Make sure the ruler is at the top of the screen again and that it measures six inches long.

2 Click **Insert**, then **Drawing**. Move the cursor to the top of the screen so that it's touching the words **Microsoft Drawing**. Hold down the mouse button, and drag the mouse down. This will move the paint screen. Move the screen down until you see the ruler on the word-processing page, and make sure the left side of the canvas is lined up with the beginning of the ruler.

3 Then click the **Rectangle** tool and draw a long rectangle that measures 4 inches, following the ruler at the top of the word-processing page. Click the **Straight-Line** tool to make the inch marks along your ruler, again checking the computer's ruler to make marks in the right places.

HINT! The reason you should write the numbers to the left of the inch marks is so that they will all fit when you print out your ruler!

4 Now click the **A icon** below the first inch mark, a little to the left of it, and type in the number 1. (If you don't see the **A icon,** click the square in the upper-right corner to make the canvas bigger.) Hit the **Space Bar** to move the cursor until it is a little to the left of the second inch mark, and type in the number 2. Keep going up to 4!

PSST!

Make sure you click the Edit command on the Paint canvas, not the word-processing page. Otherwise, you might lose your ruler because you'll be going back to the word-processing page!

5 On the paint canvas click **Edit, Select All,** then **Edit, Copy.** Exit by clicking the X in the upper-right corner of the paint canvas. You should return to the word-processing page. Click **Edit,** then **Paste.** Your ruler should appear!

6 **Save** your ruler and **Print** it out. To make your ruler more sturdy, glue it to cardboard, poster board, or construction paper. Then cut it out.

7 Now measure things around your home! On a sheet of notebook paper write down the object, then write down how long or tall it is.

8 When you've measured all the things you want to measure, compare the measurements. Write them down, showing which are greater than (**>**) and which are less than (**<**).

Example: the bookshelf (48 inches) **<** the kitchen counter (96 inches)

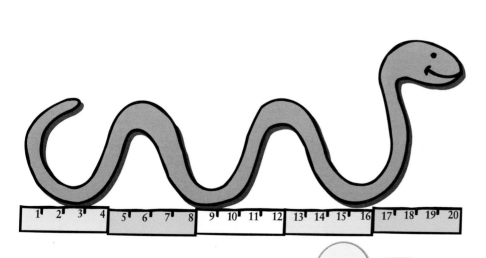

HINT! You can make your ruler longer! Make several rulers, and tape them together. Make sure you change the numbers each time, like you did for the "Sky-High" activity!

Chapter Four

Money Matters

Have you ever heard the expression "money makes the world go around"? Money isn't really the only thing—where would we be without chocolate chip cookie dough ice cream or pizza? Still, money is important, and it's what you use to buy those yummy foods! To understand how money works, you'll need some more math lessons. These activities will help you become a master of math and money.

MATH — Making Money

Now's your chance to make some money—on your computer! Make fake bills and coins, then use them to practice spending. You'll learn addition and subtraction just by spending money!

Steps:

1 Go into the **Paint** program. Start with the **Rectangle** tool, and make an outline of a rectangle. Fill it with fun colors, using the **Paint Can**. Copy and **Paste** the dollar bill so you have six of them on your screen. Make sure you move each new bill to the side after pasting it. Now fill in money amounts for each bill. Make the first one 1 dollar. Make the first one 1 dollar. Then make 5-dollar bills, 10-dollar bills, 20-dollar bills, 50-dollar bills, and 100-dollar bills. **Save** and **Print** your bills, and if you need more, just print out more copies of them later.

2 Now use the **Ellipse** tool to make coins. Look at real quarters, nickels, and dimes to remember their sizes. With the **Paint Can**, fill the coins with color. With the **Pencil** tool, write in the values for each. **Save** your money.

3 Now figure out how much each type of coin is worth. Quarters are the most—twenty-five cents. So how many dimes and nickels equal one quarter? You need two dimes and one nickel to equal a twenty-five-cent quarter. **Copy** and **Paste** the dime you painted so that you have one more dime. Use the **Dotted-Line Box** tool to move around the coins so that the two dimes are with the nickel.

4 With the **A icon**, add + and = signs between the coins to show that two dimes plus one nickel equals a quarter. Repeat these steps to show other combinations that equal a quarter, like five nickels. Or you can show how many pennies are in a nickel, or how many nickels are in a dime.

HINT! You can move any item on your page. Drag the Dotted-Line Box around it. When the thick, dark cross appears, move your picture!

5 **Save** and **Print** your coins when you're done, and keep them with the bills you made. You can use the fake money in games you play with friends!

MATH SHOP-O-RAMA

When you go to the grocery store, you have to make sure the food you buy won't cost more than the money you've brought with you. Otherwise, you'll face big-time embarrassment! Make a grocery list for things you might buy in a store, then find out the prices to see how much money you would need.

Steps:

1 In the **word-processing** program, flip through your clip-art files to find a picture of a food you would like to buy. You can also paint a picture of the food in your **Paint** program.

2 **Insert** the clip art onto the word-processing page or **Copy** and **Paste** the picture you draw in the **Paint** program onto the word-processing page.

3 Click onto the picture, then go into **Format** and click **Text Wrap**, then **Absolute**.

4 Move the cursor beside your art, and type in the food name. Also type in a line by holding down the **Shift** and the **Dash** keys (between 0 and =). Move your food name and the line with the **Tab** key or **Space Bar**.

5 Hit **Return** until the cursor is directly below the picture.

6 Repeat steps 2 through 5 to add more foods to your list.

7 Below your last food item, type in "Total: _____."

8 Now check out how much each of these foods costs by going shopping with your parents or just asking someone. Write the prices on the lines. Add up the prices to see the total amount of money you spent.

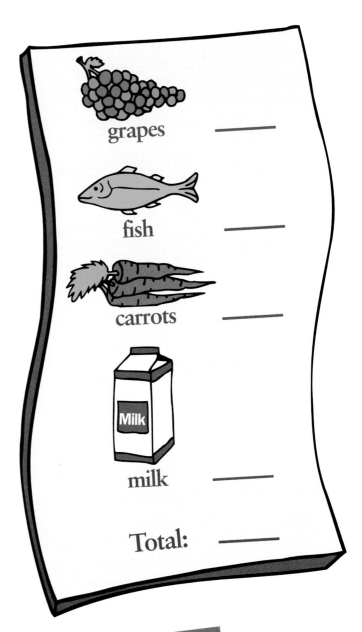

PSST!

Remember that the clip art in this book comes from sources other than Microsoft Works!

SHOPPING SPREE!

Wouldn't it be cool if there were a store you could go to and buy anything you wanted without spending any real money? Too bad that's not how the world works, huh? Just for fun, pretend your house is a big store and create some sale signs for things around your home. Then choose things to buy, add up the prices, and see how much you spent.

Steps:

1 Go into the **Paint** program. Choose the **Rectangle** tool, and create a box to make a sale sign. Fill it with a color of your choice by clicking the **Paint Can**.

2 Now draw a picture of the thing for sale. Draw the picture *outside the sign* with the **Pencil, Paintbrush, Paint Can**, and other tools. When your picture is ready, click the **Dotted-Line Box** tool. Drag a box around the picture. When you see the thick, dark cross, move your picture inside the sign.

COOL IDEA! Use your signs to set up an imaginary store! Be the store owner, and invite friends and family to stop by and shop. Let them buy things with the fake money you made on pages 26 and 27.

3 With the **Paintbrush** write in the name of the item for sale, along with its price. Make sure the **Paintbrush** is a different color from the sign color so you can see your words.

4 **Save** and **Print** your sign. Cut it out. Repeat steps 1 through 3 to make more signs. Tape a straw or pencil (or attach double-stick tape) to the back of each. Place your sale signs next to the items in your home.

5 Now go shopping! Ask a friend or someone in your family to go with you. Choose things you would like to buy, and write down the prices. After your shopping spree, add up the prices. How much money did you "spend"?

Chapter Five

SPORTS ZONE

Another area where numbers are important is sports. If you think about it, you really can't have sports *without* numbers. How would you measure speeds in a long-distance race? How would you keep score of baskets made during a championship game? How would you know how far a long jumper jumped? In the wide world of sports, numbers help us keep track of records. So if you like playing—or even just watching—you'd better learn the math.

 ## MATH: WiN, LoSE, AND DRAW

How well is your favorite team doing? How well are they doing compared to other teams in the league? The only way to find out is to compare the numbers! Make a bar graph to use for keeping track of their wins and losses.

Steps:

1 Go into the **Paint** program. Draw the outline for a bar graph the way you did in "What Do You Think?" with two straight lines going across and up and down the canvas.

2 This time write the words *Wins*, *Ties*, and *Losses* underneath the line at the bottom of the canvas. Use the **Rectangle** tool to make nine boxes above each word. Don't color in any boxes. Instead **Save** and **Print** your graph the way it is.

3 Every time your team plays, color in a box in the right column with markers or crayons or whatever you want to use. For instance, when they win, color in a box above the word *Wins*. When your team has played nine games, print out a new chart for the next nine games. Keep going until the end of the season!

4 At the end of the season count up how many boxes are colored in for each word and then note whether the team had more wins or losses.

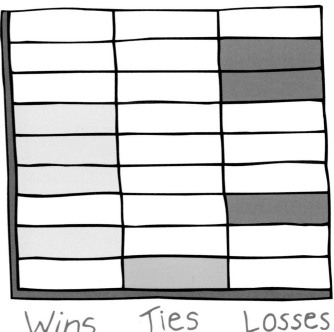

 HiGH SCORiNG

Isn't it exciting watching a game and waiting to see if your favorite team will score a basket or hit a home run? There's math again—counting all the goals to see who's winning. That's what scoreboards are for. They show you how many points each team has. Make your own scoreboard for the sport you really love to watch.

Steps:

1 Go into the **Paint** program. Choose the **Rectangle** tool, and create a long box. With the **Paint Can**, fill the box with color.

2 Now write a title on the top of the scoreboard using the **A icon**. You can move the text by using the **Space Bar** and **Tab** keys. Or move the type by dragging a **Dotted-Line Box** around it.

3 Click the **Rectangle** tool again to make small boxes for each time period in the game. These will go on your scoreboard. Make sure the boxes are small enough so that you can fit in as many as you need, making enough to record each team's score in each time period. For instance, a baseball scoreboard needs eighteen boxes because there are nine innings and you want to show what both teams score in all of the innings. A basketball game only needs eight boxes for the four quarters. Fill the boxes with white color using the **Paint Can** so that you will be able to write in them. Don't forget to make white boxes at the bottom of both columns to record the total number of points each team scored in the game.

4 Add a heading to tell which game you chose for your scoreboard.

5 Then click **File**, and check **Page Setup**. Make sure you've clicked on **Portrait**. Then **Save** and **Print**. Save out as many copies as you want, so you can use a new one at each game.

6 Use your scoreboard to keep score of your team's progress during a game. Add up all the points to show the total number scored.

BASEBALL

TEAM: _____   TEAM: _____

1st inning

2nd inning

3rd inning

4th inning

5th inning

6th inning

7th inning

8th inning

9th inning

Total